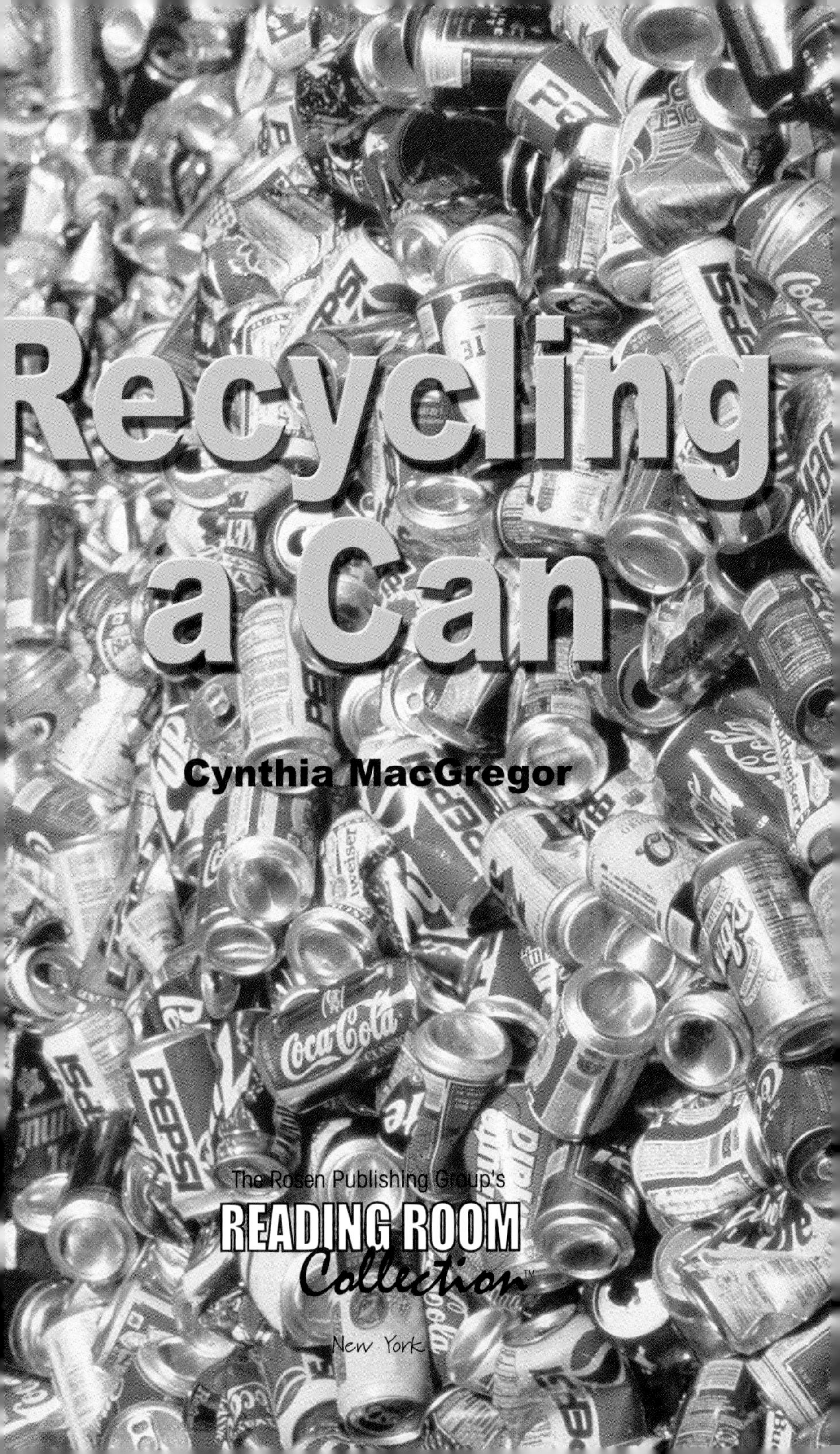

Recycling a Can

Cynthia MacGregor

The Rosen Publishing Group's
READING ROOM Collection™
New York

Published in 2003 by The Rosen Publishing Group, Inc.
29 East 21st Street, New York, NY 10010

First Library Edition 2003

Book Design: Haley Wilson

Photo Credits: Cover, pp. 1, 12 © Mitch Diamond/International Stock; p. 5 © Bokelberg/Image Bank; p. 7 © Richard Meinzer/Jeff Greenberg/SuperStock; p. 9 © Roger Holden/Index Stock; p. 10 © David Porter/Index Stock; p. 11 © Balemaster; p. 14 © Lonnie Duka/Index Stock; p. 16 © Jim McNee/Index Stock; p. 19 © James L. Amos/ Corbis; p. 21 © SuperStock/Peter Johansky/ Index Stock.

Library of Congress Cataloging-in-Publication Data

MacGregor, Cynthia.
Recycling a can / Cynthia MacGregor.
p. cm. — (The Rosen Publishing Group's reading room collection)
Includes index.
Summary: Follows an aluminum can from the time it is collected for recycling, through ninety days of different processing steps, to a new can that is ready for use.
ISBN 0-8239-3744-5 (library binding)
1. Recycling (Waste, etc.)—Juvenile literature. 2. Aluminum—Recycling—Juvenile literature. [1. Recycling (Waste, etc.) 2. Aluminum—Recycling.] I. Title. II. Series.
TD794.5 .M32 2002
673'.7228—dc21

2001007300

Manufactured in the United States of America

For More Information

Recycle City
http://www.epa.gov/recyclecity/

Reduce Recycle Reuse
http://www.leeric.lsu.edu/energy/rrr/index.html

Contents

Why Recycle?

Recycling is a way of collecting and reusing **materials**. The more we reuse old paper or old **containers**, the less trash we have taking up space in garbage dumps. We also have to cut down fewer trees to make paper, and we don't have to dig up new materials from the earth to make containers.

Soda cans are made from a material called **aluminum**. Aluminum is made from **bauxite** (BAWX-ite), which is dug up from the ground. Let's follow a recycled aluminum can and see what happens to it.

When we recycle materials, we cut down on pollution. We can recycle many things, like aluminum, glass, and paper.

The Cycle Starts

Mary and Tom live in different towns, with different recycling methods.

Tom's town has curbside recycling. This means that Tom and his family can rinse out their used cans and put them in a plastic bin. Tom and his parents put the bin out near the street. A truck comes and collects all the cans.

Mary's town has no curbside recycling, but there is a recycling center. When Mary's family has saved a lot of cans, they take the cans to the recycling center. In some states, you can get money back for your used cans.

Some people crush their cans before putting them in bags and bringing them to the curb to be picked up.

Coca-Cola

The Recycling Plant

From the recycling center or the recycling truck, the cans are sent to a recycling plant. The plant is a building full of machinery where the cans are put through the recycling **process**. The cans may get to the plant by truck, train, or ship.

It is important to recycle. In one year, we save enough energy by recycling cans to light a medium-sized city for six years! When a soda can is recycled, it isn't just refilled with more soda and sent back out again. It goes through many different steps.

A recycling plant, like the one shown here, turns aluminum cans into other things that can be used again.

ycling
DANGER
DO NOT OPERATE
WITHOUT GUARDS

Shredded and Heated

Recycling plants work in different ways. In some plants, a machine flattens the aluminum can first. Other recycling plants don't do this step. Whether or not the can has been flattened, it is **shredded** into small pieces.

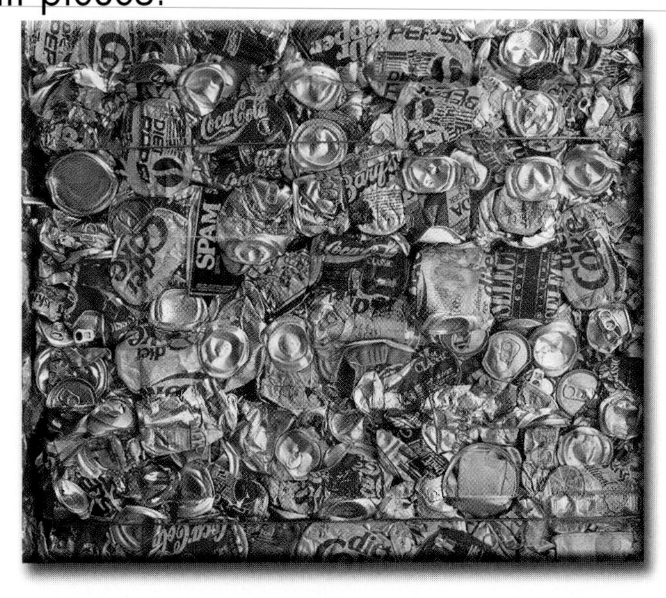

Cans like the ones above are shredded into little pieces by a machine like this one.

These aluminum pieces are put into a large oven. The oven's heat removes the paint from the aluminum. It also removes any moisture that may still be on the aluminum.

Coca-Cola
CLASSIC

Screened and Cleaned

The aluminum pieces are put on top of a large, flat screen—something like the screens that might cover your windows at home. Any dirt on the aluminum pieces falls through the screen. The clean aluminum pieces stay on top of the screen.

When the aluminum pieces are free of all paint, moisture, and dirt, they are ready for the next step in the recycling process.

Once these cans are shredded and heated to remove their paint, they must be cleaned once again before they are ready for the next step.

Into the Furnace

Next, the aluminum pieces are put into a very hot **furnace**. The furnace melts the pieces. The **molten**, or melted, aluminum is poured into **molds**. These molds are long, flat, empty containers. When the aluminum cools, it takes the shape of the molds and hardens into flat rectangles.

These big, thick sheets of aluminum are called **ingots**. Each ingot can weigh between 20,000 and 40,000 pounds. That's about ten times as much as a car weighs!

It is about 1400 degrees Fahrenheit inside the furnace. That's almost seven times as hot as boiling water!

Squashed and Flattened

A machine smooths the top and bottom surfaces of the aluminum sheets. Then the sheets, or ingots, are flattened. They are put through a machine with giant steel rollers. The rollers press the aluminum to make the sheets much thinner.

The aluminum passes between the rollers several times, and each time it comes out thinner. Finally the aluminum sheet is only about a quarter inch thick.

When the aluminum sheets are thin enough, they are ready for the next step.

A Two-Mile Coil

In another piece of machinery called a finishing mill, the aluminum is made hard enough and thin enough that it can be rolled up. The edges are trimmed, too. Now it is called a **coil**.

If you unwound one of these aluminum coils, it could be up to two miles long! It takes more than one million recycled cans to make just one of these coils.

More than 100,000 cans are recycled every minute. In ten minutes, enough cans are recycled to make one coil.

19

From Coil to Foil

Now the rolled-up aluminum coils are ready to be made into many kinds of new things. Some aluminum is used to make car parts. Some aluminum is used to make pots and pans or aluminum **foil**.

Much of the aluminum that is wrapped in coils is sent to a can **manufacturer** (man-you-FAK-chur-ur). This aluminum will be turned into soda cans once again.

Many of the things we see and use every day are made of recycled aluminum.

A Two-Month Cycle

Can manufacturers make more than ten million cans every day. A soda company will buy these cans and fill them with soda once again. There is no limit to the number of times we can recycle a can.

The recycling process takes about sixty days. That's two months from the time Tom put his can into the recycling bin until the time someone else buys a new can of soda made from the aluminum of Tom's old can. When you buy a can of soda, you may even get a can made from some of the aluminum you recycled!

Glossary

aluminum A very light, silver-white metal used to make cans, tools, car parts, and many other things.

bauxite A material found in the ground that is used to make aluminum.

coil A rolled-up sheet of recycled aluminum that is used to make new aluminum cans.

container A can, jar, box, or anything used to hold or store something else.

foil A very thin sheet of metal.

furnace An enclosed space for a very hot fire.

ingot A large, rectangular piece of metal that is ready to be turned into something else.

manufacturer A company that makes something.

material What a thing is made from.

mold A container into which you pour liquid. When the liquid hardens, it takes the shape of the container.

molten Melted.

process The act of making or doing something by following a set of ordered steps.

shred To tear or cut something into small pieces.

Index